Design and 3d printing of a negative stiffness structure

AHMED AL-KUWARI

Copyright © 2012 Ahmed Al-Kuwari

All rights reserved.

ISBN:1533411344
ISBN-13:9781533411341

Abstract

In the current project a 3D model was considered and different models were designed and developed using CATIA V5 based on the given spring model. There are 3 developed models and one modified model of the given structure, the development and design were based on various notions to improve strength and durability of the spring. These four models along with the original model were analyzed for nonlinear deflections using ANSYS Mechanical. The analysis took place using displacements as inputs and obtaining reaction forces as output. Using these as inputs the analysis was performed again. The results acquired in the form of stress and displacement has been compared for the four models along with the original one. After an in depth study and thorough analysis of the aforementioned models, it was found that the model with Hexagon structure was the best among the lot. The material for this considered model was changed and a similar analysis was carried out. The results obtained for the chosen structure for both the materials were compared. This chosen model has been 3D printed.

Contents

Abstract i

Contents ii

List of Figures iv

List of Tables vii

1   Introduction 1

    1.1   Project Aims 1

    1.2   Project Objectives 2

2   Literature Review 3

    2.1   Selective Laser Melting 3

    2.2   Fused Deposition Modelling 4

        2.2.1   Filaments 6

        2.2.2   Ultimaker 2 7

    2.3   3D Printing Industry 8

    2.4   Spring Washers 10

        2.4.1   Cylindrical Curved Washers 11

        2.4.2   Wave Washers 12

        2.4.3   Belleville Spring Washers 13

    2.5   Effect of pressures and loads 14

	2.6	Honeycomb structure 14
3	Methodology 16
	3.1	Model Design 16
	3.2	FEA Analysis 19
		3.2.1 Large Deflection Enabled 23
		3.2.2 Force Applied – Resulting Displacement 26
		3.2.3 Comparison 30
		3.2.4 316L Stainless Steel 33
		3.2.5 Reflection of Material Change 37
	3.3	3D Printing 39
4	Further Research 40
5	Conclusion 42
References 43
Appendix 49

List of Figures

Figure 1: Selective Laser Melting (Popular 3D Printers, 2016) (https://upload.wikimedia.org/wikipedia/commons/b/b2/Laser_Cladding_nozzle_configurations.jpg) 4

Figure 2: Fused Deposition Modelling (https://upload.wikimedia.org/wikipedia/commons/4/42/FDM_by_Zureks.png) 5

Figure 3: Ultimaker 2 (https://upload.wikimedia.org/wikipedia/commons/2/2c/3d_printer.jpg) 8

Figure 4: Cylindrical Curved Washer (https://upload.wikimedia.org/wikipedia/commons/b/b9/Curved_spring_washer.jpg) 11

Figure 5: Wave Washer (https://upload.wikimedia.org/wikipedia/commons/5/59/Wave_Spring.jpg) 12

Figure 6: Belleville Washers combination (https://upload.wikimedia.org/wikipedia/commons/8/85/Disc_spring_diagram.png) 13

Figure 7: Negative stiffness honeycomb with buckled beam (https://upload.wikimedia.org/wikipedia/commons/e/e

c/Horizontal_Vibration_Isolator_Beam_Column_Drawing.jpg) 15
Figure 8: Given Model 16
Figure 9: Hexagon Structure 1 17
Figure 10: Hexagon Structure 2 18
Figure 11: Honeycomb Structure 18
Figure 12: Von-Mises Stress for original model at displacement 5mm 20
Figure 13: Von-Mises stress for hexagon structure 1 at displacement 1mm 20
Figure 14: Von-Mises stress for hexagon structure 2 at displacement 2mm 21
Figure 15: Von-Mises stress for honeycomb structure at displacement 2mm 22
Figure 16: Un-converged original model at 2 mm 23
Figure 17: Von-Mises stresses in original thick model at 5 mm 24
Figure 18: Von-Mises stress results in original model for applied force 1000 N 27
Figure 19: Von-Mises stress in thickened model for applied force 1440 N 27
Figure 20: Von-Mises stress in Hexagon 1 for applied force 2350 N 28
Figure 21: Von-Mises stress in Hexagon 2 for applied force 2800 N 29

Figure 22: Von-Mises stress in Honeycomb for applied force 140 N 29
Figure 23: Force vs Displacement Graph 30
Figure 24: Force vs Maximum Von-Stress Graph....31
Figure 25: Force vs Maximum Von-Stress Stress below 40 MPa 32
Figure 26: Force vs Displacement Graph (316L SS) 34
Figure 27: Force vs Maximum Von-Stress Graph (316L SS)..35
Figure 28: Force vs Maximum Von-Stress Graph below 180 MPa (316L SS) 36
Figure 29: Force vs Max. Von-Mises stress graph for Hexagon 1 structure..37
Figure 30: 3D printed Hexagon Spring 39
Figure 31: Weighing of the model............................40

List of Tables

Table 1: Maximum von-Mises stresses of the structures at normal condition 22

Table 2: Maximum von-Mises stresses of the structures with large displacement enabled 25

Table 3: Reaction forces of the structures for the given displacement with large displacement enabled 26

Table 4: Masses of the Hexagon Structure 1 38

Table 5: Results in original model for respective applied forces 49

Table 6: Results in thickened original model for respective applied forces 50

Table 7: Results in hexagon 1 and hexagon 2 structures for respective applied forces 51

Table 8: Results in honeycomb structures for respective applied forcesThe following are the results obtained using the material ABSplus-P430 52

Table 9: Results in original model for respective applied forces (316L SS) 53

Table 10: Results in thickened original model for respective applied forces (316L SS) 54

Table 11: Results for Hexagon 1 and Hexagon 2 structure model for respective applied forces (316L SS) 55

Table 12: Results in honeycomb structure for respective applied forces (316L SS) 56

Table 13: Mass of the models 57

1 Introduction

In the process of production of a component or part there are many wastages and man hours to get a clean finished product. The production of complex structures takes more time consuming lot of resources. To overcome this additive manufacturing and 3D printing are being researched and developed.

Additive Manufacturing is the process in which the product is manufactured by using technology that deposits layer-by-layer enabling the manufacturing of complex geometry in different materials with fewer resources. It allows customization of product as per the need without much difficulty. There is numerous numbers of materials available for printing 3D objects based on the printer being used. The structural performance and stability of the product varies based on the material being used.

In this project, various sources of spring modifications which are used to optimize the performance of the spring are discussed. Different working principles of 3D printers were also studied.

1.1 Project Aims

This project aims at understanding how the 3D printing can be used in manufacturing industry and to study a spring by modifying it structure and 3D printing it.

1.2 Project Objectives

The objectives of this project are:

1. To analyse the given spring model using ANSYS and learn the problems in the model.
2. To modify the model and analyse the changes in the spring.
3. To understand the differences that are made by small structural changes.
4. To study and understand Additive manufacturing and 3D printing.
5. To learn a way to use 3D printing to manufacture the designed model.
6. To identify the effects of 3D printing on the model performance and to suggest the structures that can be printed using it.

2 Literature Review

3D printing is an Additive Manufacturing process in which a product is printed by depositing material layer-by-layer. Additive Manufacturing gives flexibility in defining the properties of the components such as strut width, pore size and density. There are different types of additive manufacturing technology. The purpose of this study is to design a negative stiffness spring using the Finite Element Method and produce it using the additive manufacturing process, (Spentzas,2002) and (Kanarachos,2008). Optimization methods are usually employed to design such structures, however in this study only a parametric analysis was conducted (Kanarachos, 2005).Furthermore, although the structural behavior beyond the yield limit is of interest, it is not part of this study (Kanarachos, 2007) (Kanarachos 2015).

The additive manufacturing technologies that were studied in this paper are:

1. Selective Laser Melting (SLM) and
2. Fused Deposition Modelling (FDM)

2.1 Selective Laser Melting

Selective Laser Melting is a process in which a high power laser is emitted onto a bed of metal powder. The laser then melts the powder in layer-by-layer process based on the geometry of the input file provided. The SLM process takes place in a controlled chamber in the presence of inert gas where the oxygen levels are less than 500 parts per million so that there would be over burning of the powder. Usually an ytterbium laser with power of 100's of watts is used for this process. Using SLM a product with internal features of size up-to 100 μm could be manufactured.

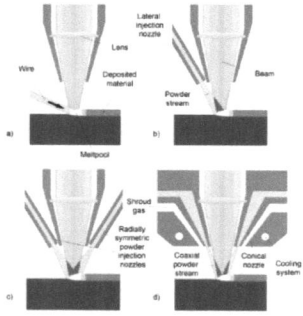

Figure 1: Selective Laser Melting (Popular 3D Printers, 2016)
(https://upload.wikimedia.org/wikipedia/commons/b/b2/Laser_Cladding_nozzle_configurations.jpg)

As shown in the figure above the scanner mirrors change the direction of the laser based on the geometry of the product. The f-θ lens concentrated the laser onto particular area intensifying the laser. The laser then reached the powder which is present on the base plate. The powder melts and forms the structure known as the melt pool. When a layer is created the build platform moves in z direction allowing space for next layer and the powder is added into the melt pool through the feed container. The extra powder which is more than the required for the layer is removed using powder scraper and is extracted from overflow container.

2.2 Fused Deposition Modelling

Fused Deposition Modelling also known as Fused Filament Fabrication (FFF) is a technique in which a plastic filament is melted and is deposited in layer on the base plate by the nozzle based on the geometry of the product. A plastic filament or metal wire is supplied to the printer head where it is heated and melted and printed in layers by using nozzle at a controlled rate based on the product being printed. FDM has ability to print up-to 9 different colors. These colors can be printed on the same model as well. The drawback of a FDM is it leaves few hanging on the model and cannot produce product of some geometries. It needs to have the product oriented in such a way that it has supportive base being printed first or a removable support needs to be placed. Contrary, in SLM a product can be manufactured by keeping it in any orientation.

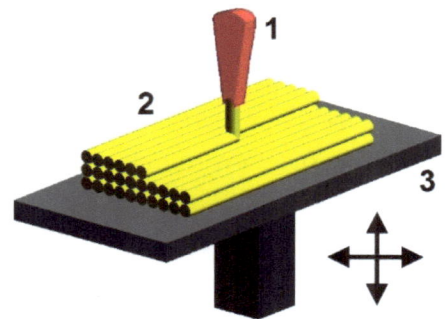

Figure 2: Fused Deposition Modelling (https://upload.wikimedia.org/wikipedia/commons/4/42/FDM_by_Zureks.png)

The above figure shows the process of FDM. The filament from the filament spool is lead to the extruder to feed it to the printer head. The extruder controls the feed amount of filament to the printer head based on the requirement. When the filament reaches printer head it is heated and melted and is printed in layers by position the nozzle where required. In some models fans are attached to the printer head to cool down the printed material making it ready for the next layer to be printed on.

2.2.1 Filaments

Like the ink cartridges in inkjet printers, the filaments are the cartridges for the 3D printers. They are available in different colors based on the requirement and are easy-to melt. These filaments are fed through the printers head and heated up and extruded through nozzle for printing. At present there are many materials available in market while the filaments Acrylonitrile Butadiene Styrene (ABS) and Polylactic Acid (PLA) are prominent. These are thermoplastics which are soft and easily moldable when heated and return back to solid after cooling down. The printed object could be reused by melting without any issues.

Acrylonitrile Butadiene Styrene (ABS)

Acrylonitrile Butadiene Styrene is a polymer and in general it has strong plastic with mild flexibility. This flexibility is ABS makes it easy to work with interlocking pieces and is easier to recycle. ABS has an odor of hot plastic while heated to particular temperatures. ABS has difficulty in printing sharp edges and prints slightly smooth rounded edges due to the curling of the molten material between the nozzle and the printed layer. This could be eliminated by an extent with the use of controlled active cooling. ABS is a preferred material in mechanical use due to its flexibility, strength, high temperature resistance and machinability. (Chilson, 2013)

Polylactic Acid (PLA)

Polylactic Acid is produced from number of plant products such as potatoes, corn or sugar-beets making it earth friendly plastic compound. PLA is rigid which makes it difficult for working with interlocking pieces like ABS. PLA being made from plant products gives a smell similar to semi-sweet cooking oil. Unlike ABS, PLA doesn't have curling and sharper detail could be achieved when actively cooled. The strength of the printed object could be increased by increasing the flow which allows for stronger binding between the layers. PLA is not preferred for some structures due to its low melting temperature. The temperature is low that it the printed part could melt when kept in a hot car for a day. (Chilson, 2013)

2.2.2 Ultimaker 2

In this project for printing the designed model Ultimaker 2 was used. It is a 3D printer that used FDM technology to print the products. Ultimaker uses ABS and PLA materials and also has ability to print using other thermoplastic materials such as Acrylic and Nylon. As it is a single extruder machine it could only use single material at a time and could print only single color at a time. Ultimaker 2 printer head consists of two fans each on either side that help in cooling the printed material.

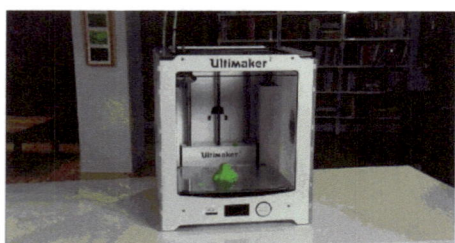

Figure 3: Ultimaker 2 (https://upload.wikimedia.org/wikipedia/commons/2/2c/3d_printer.jpg)

Ultimaker 2 has ability to print at a maximum speed of

300mm/s and with accuracy of 0.02mm layers. The Ultimaker 2 could a build a maximum volume of 23 x 22.5 x 20.5 cm with a precision of 12.5 microns in x and y-direction and 5 microns in z-direction with a nozzle of diameter 0.4mm. In Ultimaker 2 the filament is heated to a temperature of 180^0-260^0 C at the printer head to melt the filament and print it. (3D Print Works, 2014)

2.3 3D Printing Industry

3D printing was present since 1980's but it isn't famous till couple of years ago. As per the studies by (Lentejas Jr. 2014) USA constitutes 38% of the worldwide 3D printing industry followed by Japan with 9.7%. Germany with 9.4%, China with 8.7% and United Kingdom with 4.2$

3D printing has the potential to become a global revolution in the field of production technology. It has immense scope and currently has lot of impact on the Aerospace industry and the other manufacturing spheres would soon witness a paradigm shift because of this technology. The evolution of 3D printing is coined as the Third industrial revolution. The lead time for manufacturing certain products would drastically go down because of this 3D printing. There is continuous evolution in 3D printing from making prototypes to manufacture finished goods. 3D printing has also shown a great effect on the ammunition manufacturing industry. It plays a major role in the International Political Economy, currently China which is considered as the manufacturing hub, the manufacturing part can be taken away from here and be placed in the country of consumption. This also helps in reducing the carbon footprint. The following four countries are growing big in terms of 3D Printing. China looking to expand its 3D printing set up from 8% currently to a great extent, South Korea started creating their own 3D printers, New Zealand is considering 3D printing as Blueprint of the Future and South Africa which has companies trying to make 3D printers at affordable prices. These are from the developing countries and the developed countries are moving at a rapid pace in 3D printing technology.

Based on the report by (Wohlers 2014) the usage of the additive manufacturing in major areas in industry is estimated as 18.5% in Industrial/Business Machines, 18% in Consumer products/electronics, 17.3% in Motor vehicles, 13.7% in Medical sector and 12.3% in Aerospace.

Rolls-Royce has made a jet engine fitted with the largest component made using Additive Laser Manufacturing 3D printing. The component made is a bearing of 150cm diameter and 50cm thick for one of its XWB engines (Tovey 2015)

General Electric (GE) recently tested its GE9X jet engine which is the world's largest jet engine at the moment and it contains 19 fuel nozzles which are 3D printed. It is stated that the weight of the nozzle was reduced by around 25% with the use of Additive Manufacturing (Benedict. 2016)

2.4 Spring Washers

Spring washers are small in size and are used to provide tension in bolted assemblies. Based on the structure the spring washers are basically divides into three types:

1. Cylindrical curved washers
2. Wave washers
3. Belleville washers

When the assemblies are bolted using the spring washers the force is applied on the washers and the washers tend to push back due to their nature. This keeps the parts stable and prevents the nuts and screws for loosening. These washers also act as vibration absorbers in some cases.

Spring washers are usually made of steel with a combination of spring brass, phosphor bronze and beryllium copper etc. To prevent from corrosion the washers are coated using techniques such as electrogalplating, cadmium plating, electro-plating, nickel and chromium plating, black oxide, etc.

2.4.1 Cylindrical Curved Washers

These washers are used for light loads where the loads are applied repeatedly. The height of these washers is recommended to be in the limit that is less than half of their outer diameter. The following image shows the cylindrical curved washer and its cross section.

Figure 4: Cylindrical Curved Washer (https://upload.wikimedia.org/wikipedia/commons/b/b9/Curved_spring_washer.jpg)

To the load that could be taken by the cylindrical curved spring washers is calculated by the following formula:

$P = 4 \cdot E \cdot f \cdot b \cdot t^3 / D^3 \cdot K$

Where P is the applied load

    E is the young's modulus

    f is the deflection in the washer

    b is the radial width of the washer

    t is the thickness of the washer

    D is the outer diameter of the washer

    K is the stress correction factor

2.4.2 Wave Washers

These washers are used for the static loads ranging from light to moderate for up-to hundreds of pounds. For the best performance of this washer it is recommended that the ratio of mean diameter to the radial width of the washer to be equal to 8 i.e $\frac{D_m}{b} = 8$. The washer should have at least three waves for it to perform well. As the number of waves is increased the thickness of the washer has to be reduced gradually for flexibility of the washer for the specified load.

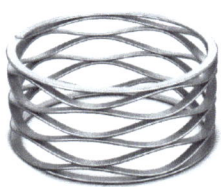

Figure 5: Wave Washer (https://upload.wikimedia.org/wikipedia/commons/5/59/Wave_Spring.jpg)

2.4.3 Belleville Spring Washers

These washers are used for high loads and are capable of taking up to 14,000 kg/cm2 loads. There are few guidelines that govern the Belleville washers which are:

1. The height to the width ratio of the rim should be below 1:10
2. The thickness to the width ratio should be around 1:5 and not exceeding 1:10

3. The ratio of outer diameter to the inner diameter should be in the range 1.5 to 1.7

The height to the thickness ratio is suggested to be maintained below 0.8 for the washer to perform well as there are chances of snapping if the h/t ratio is greater than 0.8.

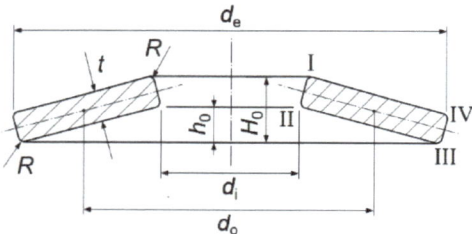

Figure 6: Belleville Washers combination (https://upload.wikimedia.org/wikipedia/commons/8/85/Disc_spring_diagram.png)

For improved performance these washers can be combined in series, parallel and series-parallel. In series combination the maximum load that could be taken by the whole stack is equal to the maximum load taken by a single washer and the deflection of whole stack is n times the deflection of each washer where n is the number of washers. In parallel combination the load that could be taken and the deflection are vice-versa to that of series combination.

2.5 Effect of pressures and loads

In the paper by (Stratis & George, 2006) composite materials such as metal reinforced plastic pipes are explored and tests were conducted on it. High Density Polyethylene (HDPE) is selected as it satisfies the required constraints for the current needs of the pipes and the tests of HDPE are compared with the results of steel. Both the tubes are subjected to different internal pressures, external loads and external loads at different pressures and are compared. It was observed that the internal pressure has more effect on the pipe at high pressures while the loads effect more at low internal pressures.

This explains that the effect of a load or force on a component depends on the other loads, forces or constraints acting on the component.

2.6 Honeycomb structure

In a paper by (Correa, et al., 2015) the honeycomb structure is studied using different materials and minor shape changes. The performance of a negative stiffness honeycomb is compared with a regular honeycomb which is both manufactured by 3D printer using adaptive manufacturing. The negative stiffness beams exhibit high initial stiffness and are good at shock absorption. Negative stiffness honeycomb is manufactured using nylon 11 material in which a central beam is used as shown in the following figure to prevent the honeycomb from rotation and lateral expansion of vertical walls. It also helps in eliminating the requirement for supports.

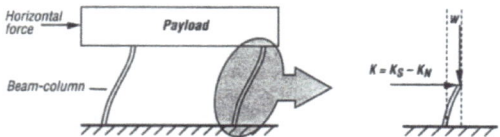

Figure 7: Negative stiffness honeycomb with buckled beam (https://upload.wikimedia.org/wikipedia/commons/e/ec/Horizontal_Vibration_Isolator_Beam_Column_Drawing.jpg)

From the comparison of experimental and FEA results it was observed that there is quite difference in the reaction forces exhibited by the structure. These might be due to properties of nylon 11 material, plastic deformation of the structure and deformations in rigid beams.

The regular honeycomb is also tested and it is observed that the honeycomb structure behaves similarly under loading conditions but the regular honeycomb couldn't recover to original shape after the compression.

3 Methodology

The model that is taken for this project is a spring as shown in the figure below.

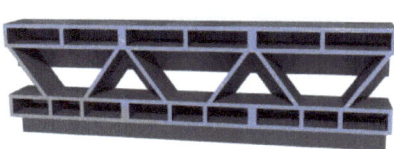

Figure 8: Given Model

The model was measured and its dimensions are 91.351(b) x 24.287(d) x 10(h) mm. The thickness of the beams in the spring is found out to be 1 mm each while the beams that are thick are 2mm. The height of the center part (spring) is 12.287 mm. The mass of this model is calculated using the density of the material that would be used to print the spring (i.e., 1040kg/m^3) and is found out to be 561gm. After the initial inspections are done the model is exported to IGES for use in ANSYS.

3.1 Model Design

After the initial model was seen and a brief analysis was done 3 ideas were taken in process of improving the strength of the spring. These ideas of the spring are taken from the (Correa, et al., 2015) and are improvised accordingly.

Initially the center part of the model is removed and hexagons are placed instead of the beams as shown in the figure below:

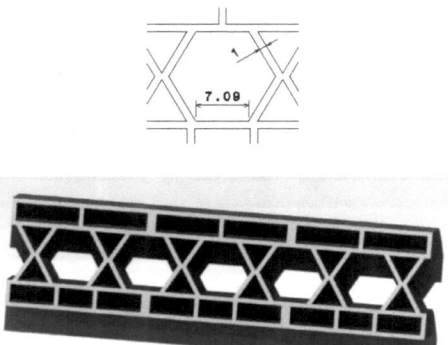

Figure 9: Hexagon Structure 1

Later the center part of the beam s modified by rotating the hexagons by 90⁰ and altering the dimensions accordingly as shown in the figure below

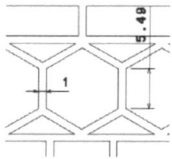

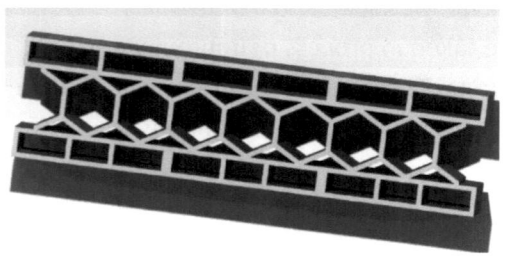

Figure 10: Hexagon Structure 2

The center part is altered to form a honeycomb structure

with the dimensions as shown in the figure below and the resulting is shown in the figure below

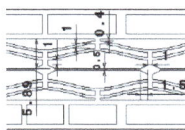

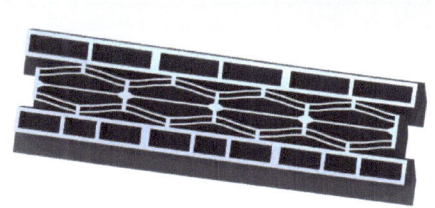

Figure 11: Honeycomb Structure

These structures are then used for analyzing and selecting the best design using FEA software ANSYS.

3.2 FEA Analysis

The given model is analyzed in ANSYS Mechanical to find out the force that needs to be applied on the 3D printed model for experimental values to get a deflection of 1-5mm.

Initially the model is imported into ANSYS Workbench with Static Structural as solver and the material properties are given as follows (Stratsys, 2015):

Density $1040 \ kg/m^3$

The Isotropic Elasticity values are derived from young's modulus
Young's Modulus 2320 MPa
Poisson's Ratio 0.35

Bilinear Isotropic Hardening values
Yield Strength 37 MPa
Tangent Modulus 2100 MPa

Later the model is meshed with a fine 3mm mesh and the bottom of the model is fixed. A varying displacement is given to the top of the model and this displacement is modified as per the requirement as shown in the table 1 and analysis is run with non-linear properties enabled. The Von-Mises stresses for the given model at a displacement of 5mm are shown in the figure below.

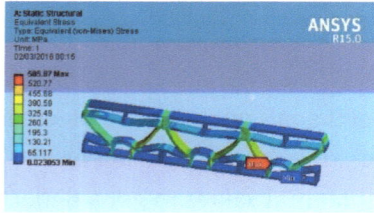

Figure 12: Von-Mises Stress for original model at displacement 5mm

From the figure it can be observed that the maximum stresses for the model are near the joint of the bottom part and the center part. In an attempt to reduce the stresses the model is modified with a hexagonal spring part and honeycomb structure in the center. This idea of hexagonal spring is taken from (Correa, et al., 2015).

The analysis is run for the hexagon model and the results of Von-Mises stress are shown in the figure below at displacement 1mm. It can be observed from the figure that the maximum stress has been reduced by very small amount (refer table no.1) and the stress has also been distributed well along the structure decreasing chance of cracks at corners.

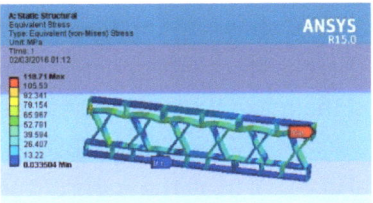

Figure 13: Von-Mises stress for hexagon structure 1 at displacement 1mm

Now the orientation of the hexagon was rotated to observe the change in distribution of stress in the structure. The hexagons are rotated by 90^0 and the analysis is rerun for the new model and the results of Von-Mises stress at a displacement of 2 mm are shown in the figure. From the results it can be observed that the maximum von-Mises stresses in the model are greatly reduced (refer table no.1) and also the stresses are well distributed in the structure.

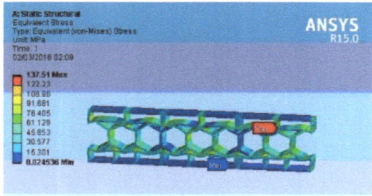

Figure 14: Von-Mises stress for hexagon structure 2 at displacement 2mm

The analysis is also done with honeycomb structure to find out the behavior of spring with honeycomb structure in the center part. The result of stresses with displacement 2mm is shown in the figure below. It is observed that there is lot of difference in the stress distribution and the maximum von-Mises stress is 60 MPa while it is 137 MPa for the hexagon structure. The stress distribution is also equal in all the parts of the honeycomb structure.

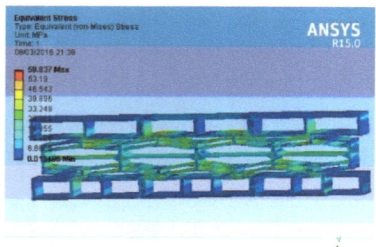

Figure 15: Von-Mises stress for honeycomb structure at displacement 2mm

After obtaining the results of all the structures at different input displacements the values are tabulated in the following table:

Displacement (mm)	Max von-Mises Stress (MPa)			
	Original Model	Hexagon 1	Hexagon 2	Honeycomb
1	120.52	118.71	70.734	35.14
2	236.58	232.71	137.51	59.84
3	352.95	345.97	203.84	68.30
4	469.40	458.97	270.28	89.94
5	585.87	571.81	336.74	111.5

Table 1: Maximum von-Mises stresses of the structures at normal condition

From the results it is clear that even though the structure being hexagon its orientation is really important and does show a great impact in the working of spring. It can also be said that the Honeycomb structure would undergo into less internal stresses at a displacement of 5mm when compared to the other models.

3.2.1 Large Deflection Enabled

Now with the large displacement option enabled, the analysis for all the four models is run. Unlike the normal analysis, with the large deflection enabled the original model could take only up-to an input displacement of 1.9138 mm and the model couldn't converge making the model unable to take a force that could displace it more than 2mm resulting as shown in the figure below.

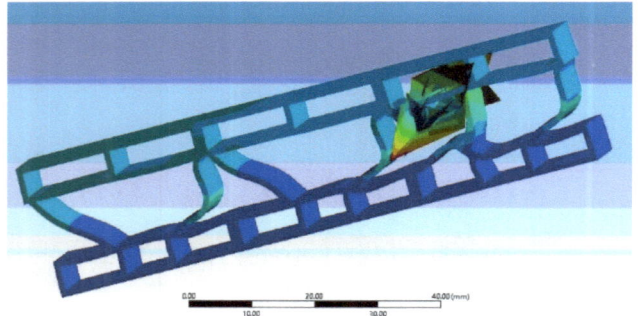

Figure 16: Un-converged original model at 2 mm

As the original model has failed to converge the thickness of the center structure of the model is increased by 0.5mm and is verified for convergence.

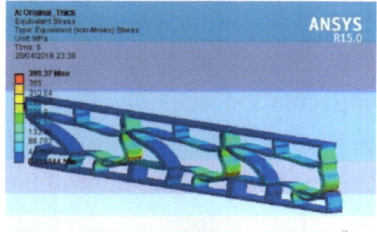

Figure 17: Von-Mises stresses in original thick model at 5 mm

Unlike the original model the thickened model could take input displacement of up-to 5 mm. As seen above the stresses are in one side part of the spring while the beams in other direction show very less stresses. The maximum stress for an input displacement of 5mm is 399.37 MPa. There hasn't been much change in distribution of stress in the structures of the other three models.

The results for the four models Thickened Original, Hexagon 1, Hexagon 2 and Honeycomb structures are tabulated below for respective input displacements:

Displacement (mm)	Max von-Mises Stress (MPa)			
	Thickened Original	Hexagon 1	Hexagon 2	Honeycomb
1	125.24	120.62	69.225	35.892
2	221.32	201.84	135.35	62.004
3	278.4	270.37	174.77	64.698
4	344.26	298.30	203.00	69.212
5	399.37	327.73	236.27	69.693

Table 2: Maximum von-Mises stresses of the structures with large displacement enabled

From the results it can be observed that there isn't much difference in the maximum von-Mises stresses for 1m and 2mm displacements compared to the results without large displacement option and from there has been a huge difference in the maximum von-Mises stresses.

Along with the von-Mises stresses the reaction forces in the fixed end are also calculated in the ANSYS to estimate the forces that need to be applied on the model. These values are tabulated in the table below:

Displacement (mm)	Reaction Force (N)			
	Thickened Original	Hexagon 1	Hexagon 2	Honeycomb
1	3020.7	2762.7	1771.6	118.84
2	3174.8	4698.2	2752.9	106.14
3	3018.5	5315.3	2810.0	78.566
4	3074.3	5480.8	2891.3	98.759
5	3044.8	5542.7	2984.2	76.293

Table 3: Reaction forces of the structures for the given displacement with large displacement enabled

3.2.2 Force Applied – Resulting Displacement

With the reaction force results in the table no.3 as input the forces are applied on the structures to obtain the deflections for the applied forces with large deflection option enabled. While these reaction forces when given as input gave different deflection than expected. In a trail to find the maximum force the structures could take, the input force is modified and is increased slowly.

From the FEA analysis results it is observed that the original model could take up-to a maximum force of 1164.1 N, while the thickened original model could take a maximum of 1667.3 N, hexagon structure 1 could take a maximum of 3151.4 N, hexagon 2 structure could take a maximum of 2819.7 N and the honeycomb structure could take a maximum force of 142.31 N. It can be seen clearly that the hexagon 1 structure could take more force compared to others but the deformation of the model should also be considered before we conclude that the model is best in these five models.

As the maximum force the structures could take is obtained the forces a bit less than that are applied to the structures so that the analysis would not give abnormal results and the results are noted down for comparison.

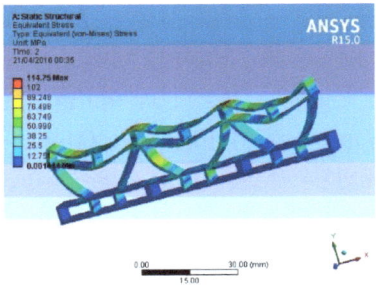

Figure 18: Von-Mises stress results in original model for applied force 1000 N

As the maximum force it could take is 1164.1 N a force of 1000 N is applied and the results of the von-Mises stress are shown above. At this force the maximum von-Mises stress is 114.75 MPa and the maximum deformation is 4.34mm. As shown in the above figure the deformation is more in the top part of the spring and some deformation could be seen in the center part of the spring.

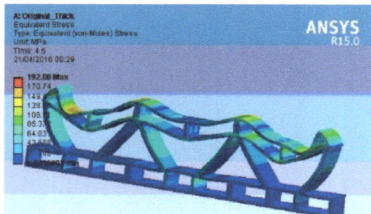

Figure 19: Von-Mises stress in thickened model for applied force 1440 N

As the maximum force it could take is 1667.3 N a force of 1600 N is applied. At the force of 1600 N it is observed that the top part is interlinking and hence a weight of 1440 N is selected for observation. The above figure shows the von-Mises stress results in thickened original model for an applied force of 1440 N. At this force the maximum von-Mises stress is 192 MPa and the maximum deformation is 7.36 mm. As shown in the above figure the deformation is more in the top part of the spring and the interlinking could be observed and very less deformation could be seen in the center part of the spring.

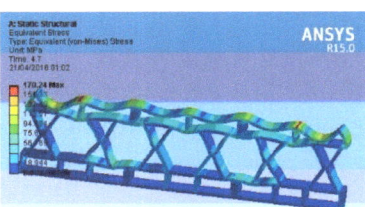

Figure 20: Von-Mises stress in Hexagon 1 for applied force 2350 N

As the maximum force it could take is 3151.4 N a force of 3000 N is applied. At the force of 3000 N it is observed that the top part is interlinking and hence a weight of 2350 N is selected for observation. The results of von-Mises stress are shown in the above figure and the maximum von-Mises stress is 170.24 MPa and the maximum deformation is 4.39 mm. As shown in the above figure the deformation is more in the top part of the spring and the interlinking could be observed and very less deformation could be seen in the center part of the spring.

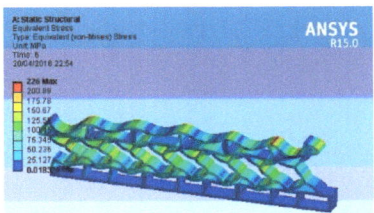

Figure 21: Von-Mises stress in Hexagon 2 for applied force 2800 N

The maximum force this structure could take is 2819.7 N and hence a force of 2800 N is applied and its von-Mises stress results are shown in the above figure. At this force the maximum von-Mises stress is 226 MPa and the maximum deformation is 10.56 mm. As shown in the above figure the deformation is more in the top part and center part and as it can be seen clearly the whole part bends towards a side.

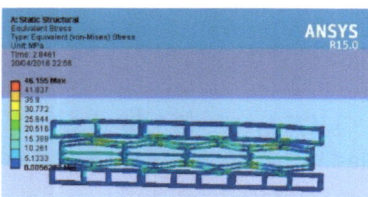

Figure 22: Von-Mises stress in Honeycomb for applied force 140 N

The maximum force this structure could take is 142.31 N and hence a force of 140 N is applied and its von-Mises stress results are shown in the above figure. At this force the maximum von-Mises stress is 46.15 MPa and the maximum deformation is 2.08 mm. As shown in the above figure the deformation towards the bottom surface and is proportional over the surface.

3.2.3 Comparison

The displacement results of all the structures are noted a graph is drawn for easy understanding and comparison. The graph is shown below:

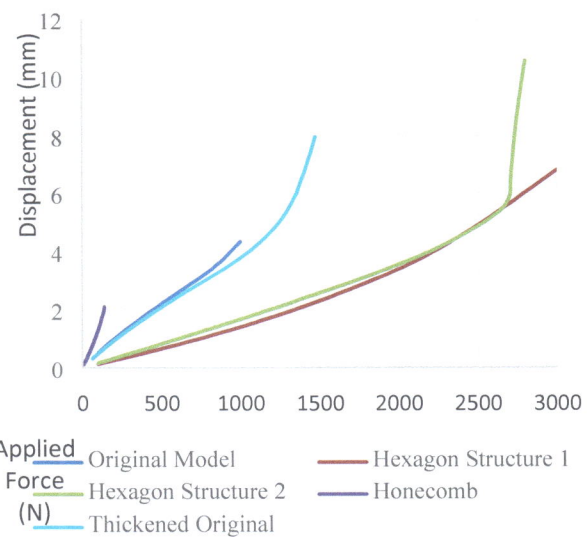

Figure 23: Force vs Displacement Graph

From the graph it is clear that honeycomb structure fails quickly at less force and the Hexagon structures could take more force with less deflection. And after a force of 2700N the hexagon 2 structure tends to deflect more when compared to the hexagon structure 1 which maintains the same deflection proportion until its maximum 3151 N.

Similarly the maximum von-Mises stress results are also drawn in graph for more understanding on the structures which are shown below:

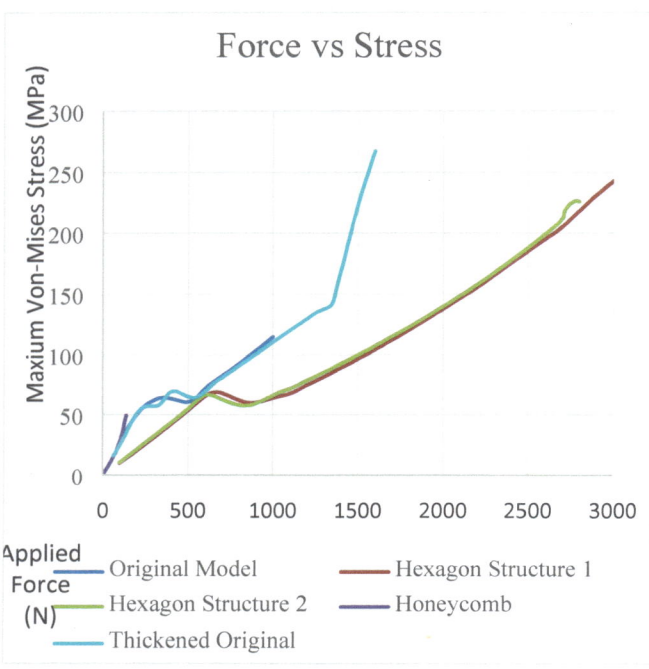

Figure 24: Force vs Maximum Von-Stress Graph

From the Force vs Stress graph it can be observed that the hexagon structures could take mores forces while having less stresses in the model. As the yield strength of the material ABSplus-P430 is 37 MPa only and hence any force that has stress more than 37 MPa would deform the structure and wouldn't be able to return to its shape making it a failure. So a close graph is taken for stresses above 37 MPa and is shown below

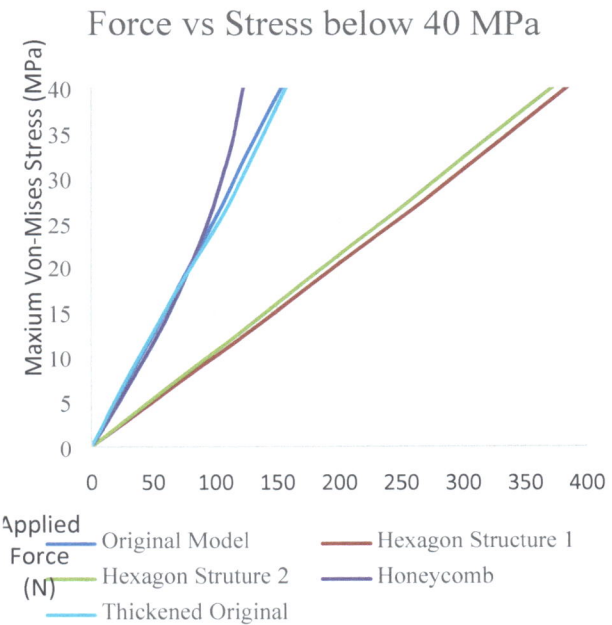

Figure 25: Force vs Maximum Von-Stress Stress below 40 MPa

From the above graph it is clear that even for max Von-Mises stress of 37 MPa the hexagon structures could take more than 300 N of force. And it is also clear that the hexagon structure 1 could take a bit more force than the hexagon structure 2 before reaching stress value 37 MPa.

3.2.4 316L Stainless Steel

The model is analyzed using ABS material. To find more about the performance of the same structure for another property another set of analysis is made. As described in paper by (R, et al., 2014) the structures are printed in 316L Stainless Steel and we are also aware that most of the springs are made using Stainless Steel. Hence 316L stainless steel is selected for analysis of the next set. The properties of the 316L Stainless Steel are as follows (AZoMaterials, 2016):

Density $8000\ kg/m^3$

The Isotropic Elasticity values are derived from young's modulus
Young's Modulus 193 GA
Poisson's Ratio 0.3

Bilinear Isotropic Hardening values
Yield Strength, Ultimate 170 MPa
Tensile Strength, Ultimate 485 MPa

The structures have been analyzed with the same forces stated earlier as it would make it good for comparison of the changes in the structure. The forces are 1000N for Original Model, 1600N for Thickened Original Model, 3000N for Hexagon Structure 1, 2800N for Hexagon Structure 2, and 140N for Honeycomb Structure. The results obtained are shown in Appendix and the graphs are drawn for the values. The following graph shows the maximum displacement in the structures for the applied forces.

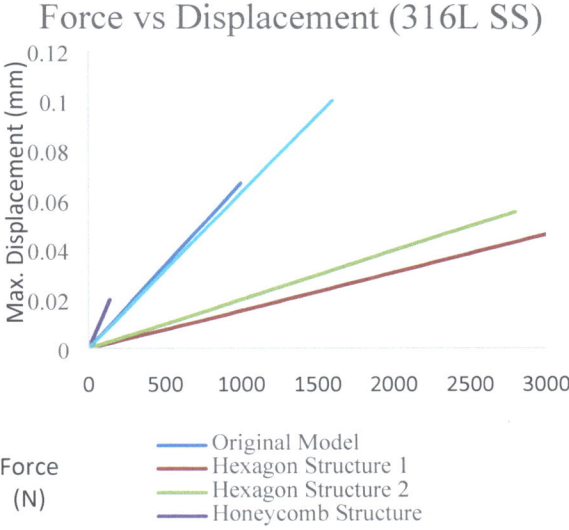

Figure 26: Force vs Displacement Graph (316L SS)

Similar to the graph for ABSplus-P430 material the trend of the graph for 316L Stainless Steel also shows that the hexagon structures perform well and the hexagon structure 1 has less deformation than any other structure at any point of load.

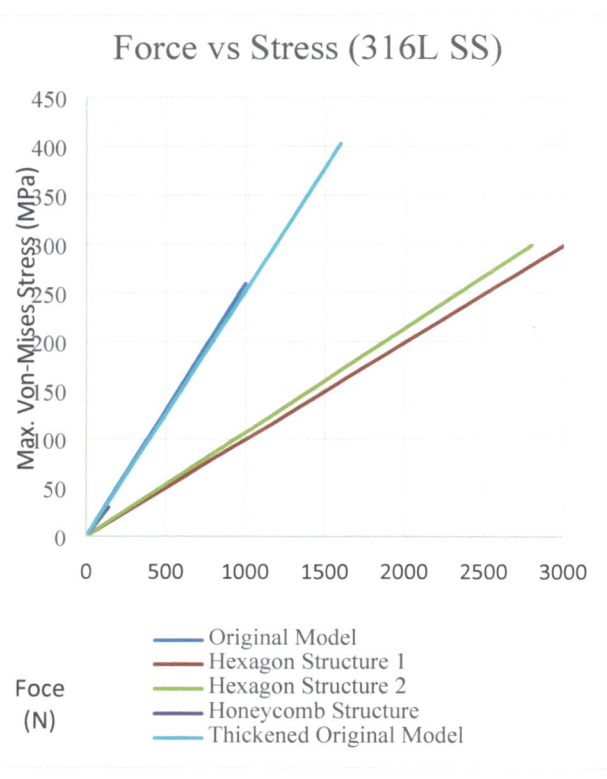

Figure 27: Force vs Maximum Von-Stress Graph (316L SS)

The above graph shows the Force vs maximum Von-Mises stress results for different structure for the material 316L SS. The trend of this graph is also similar to the graph of ABSplus-P430 material except that this graph is maintaining a constant proportion while the ABSplus-P430 graph isn't. From this graph also it is clear that hexagon structure 1 performs well at any point of force causing less stresses in the component.

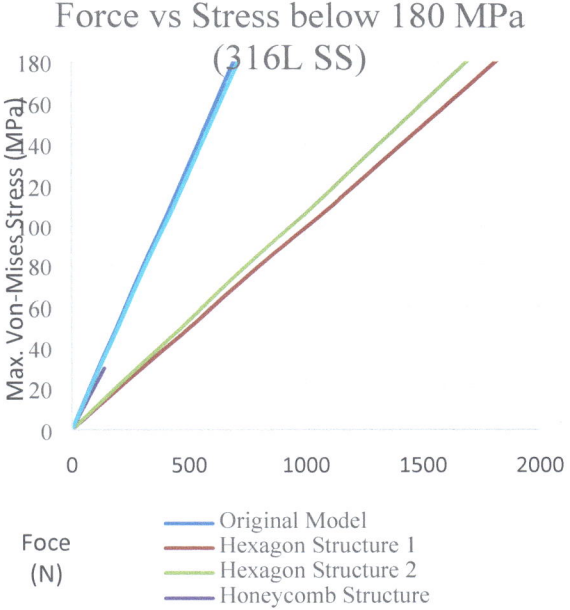

Figure 28: Force vs Maximum Von-Stress Graph below 180 MPa (316L SS)

As the yield strength of 316L stainless steel is 170 MPa, any stresses more than the yield strength would deform the structure into unrecoverable shape and hence a graph showing stresses below 180 MPa is made and is shown in the figure above. From the figure it can be observed that the Original shaped models wouldn't take force more than 700N without exceeding the yield strength. The hexagon structures could take more than 700N and clearly hexagon structure 1 could take around 1700N while the hexagon structure could take around 1600N only making the hexagon 1 structure to withstand more force.

3.2.5 Reflection of Material Change

As it is clear that the hexagon structure 1 is best structure in tests for two different materials. So the results for different materials of the hexagon structure 1 are compared to see what change was brought in the structure due to the change of material.

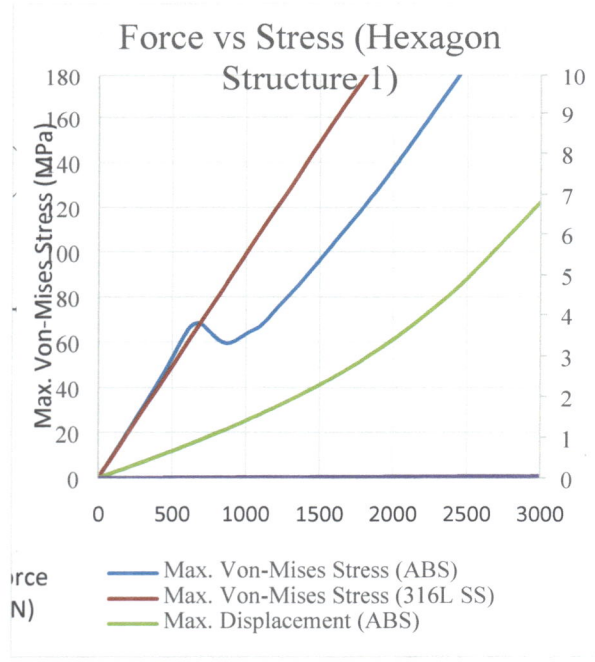

Figure 29: Force vs Max. Von-Mises stress graph for Hexagon 1 structure

It is clear from the graph that for the applied force there isn't much change in the stress vales and the stress values for the material ABSplus-P430 (blue line) are less for the applied forces than that of the stress values for the material 316L SS (red line). So it can be said that the structure with ABSplus-P430 material could take more forces than 316L SS material structure but due to the yield strength this is limited.

There is a lot of change in the displacement due to change in material. The displacement for the ABSplus-P430 material (grey line) is lot more than the displacement of the 316L Stainless Steel structure (orange line) almost by 100 times. So in this condition material 316L SS is more recommended than the ABSplus-P430. To confirm this the mass of the hexagon structure 1 is also calculated and is tableted below:

Model	Mass (ABSplus-P430) (gm)	Mass (316L SS) (gm)
Hexagon Structure 1	6.43	49.47

Table 4: Masses of the Hexagon Structure 1

The mass of the 316L SS is too high compared to the mass of the ABSplus-P430 model and is almost 7.69 times greater. This is due to the change in the density of the material. This increase in density increases the stiffness of the structure which might make the structure unsuitable for the work it I intend for. The exact usage of this spring plays a main role for selection of the material. Usage of 316L SS would increase the weight of the model resulting in increase in total weight of the product where the spring would be used. But before using ABSplus-P430 it is to be made sure that the forces are below 350N or else it would result in unrecoverable deformed shape and the spring has to be replaced every time. So it is clear that the hexagon structure 1 made with ABSplus-P430 could be used in places only where the weight on the spring is not more than 35.5 kg.

The value show above are obtained by using Measure option in CATIA and giving the density of the materials i.e. 1040 kg/m³ and 8000 kg/m³ for the materials ABSplus-P430 and 316L Stainless Steel respectively.

3.3 3D Printing

As it is clear that hexagon 1 structure is best from all the above comparison the structure is 3D printed using Ultimaker 2 3D printer.

Figure 30: 3D printed Hexagon Spring

The above figure shows the 3D printed hexagon spring. After the model is printed to make sure the properties of the spring are equal to the properties taken in analysis the spring is weighted and is found out to be 6gm which is shown in figure below

Figure 31: Weighing of the model

As the weighing machine measure a minimum of 1gm and not in milligrams the exact weight of the model could not be obtained. But it is equal to the weight calculated from CATIA for this model i.e. 6.43gm so it can be expected that the material considered for analysis in ANSYS is also correct and the results are of this product.

4 Further Research

Due to lack of the equipment further work which is experimental test of the 3D printed model could not be carried out. So if any further research would be carried out on this project the model could be tested using the forces from the analysis and the changes in the structure could be observed. And the analysis results could be cross checked with that of the experimental results.

5 Conclusion

The analysis for the different structure is carried out at different forces and found out that the initial hexagon structure is best suitable for usage than other models and could take a maximum stress 3000N but would lose its stability to recover to its original shape after the force of 350N as its yield strength is 37 MPa. So it could take a maximum weight 35.5kg without losing its ability to recover but beyond that would make is unrecoverable.

References

3D Print Works. (2014) *Ultimaker 2 3D Printer* [online] available from <http://www.3d-print-works.com/ultimaker-2-3d-printer/> [18 April 2016]

AZoMaterials. (2016) *Stainless Steel - Grade 316L - Properties, Fabrication And Applications (UNS S31603)* [online] available from <http://www.azom.com/article.aspx?ArticleID=2382> [20 April 2016]

Benedict. (2016) *GE Fires Up World's Largest Jet Engine, 3D Printed Fuel Nozzles Put Through Paces* [online] available from <http://www.3ders.org/articles/20160425-ge-fires-up-worlds-largest-jet-engine-3d-printed-fuel-nozzles-put-through-paces.html> [25 April 2016]

Chilson, L. (2013) *The Difference Between ABS And PLA For 3D Printing* [online] available from <http://www.protoparadigm.com/news-updates/the-difference-between-abs-and-pla-for-3d-printing/> [19 April 2016]

Correa, D., Klatt, T., Cortes, S., Haberman, M., Kovar, D. and Seepersad, C. (2015) "Negative

Stiffness Honeycombs For Recoverable Shock Isolation". *Rapid Prototyping Journal* 21 (2), 193-200

ISO. (2015) *Manufacturing Our 3D Future (2015-05-05) - ISO* [online] available from <http://www.iso.org/iso/news.htm?refid=Ref1956> [25 April 2016]

John Manners-Bell & Ken Lyon, S. (2016) *The Implications Of 3D Printing For The Global Logistics Industry - Supply Chain 24/7* [online] available from <http://www.supplychain247.com/article/the_implications_of_3d_printing_for_the_global_logistics_industry> [25 April 2016]

Kanarachos, S., Mathew, J., Chroneos, A., Fitzpatrick, M. Anomaly detection in time series data using a combination of wavelets, neural networks and Hilbert transform (2015) IISA 2015 - 6th International Conference on Information, Intelligence, Systems and Applications, art. no. 7388055,

Kanarachos, S. Analysis of 2D flexible mechanisms using linear finite elements and incremental techniques (2008) Computational Mechanics, 42 (1), pp. 107-117.

Kanarachos, S., Demosthenous, G. Modeling the mechanical behavior of composite metal plastic pipes subject to internal pressure and external soil and traffic loads (2007) WSEAS Transactions on Systems, 6 (5), pp. 908-913.

Spentzas, K.N., Kanarachos, S.A. An incremental finite element analysis of mechanisms and robots (2002) Forschung im Ingenieurwesen/Engineering Research, 67 (5), pp. 209-219.

Kanarachos, S.A., Koulocheris, D.V., Spentzas, K.N.Synthesis of nonlinear dynamic systems using parameter optimization methods A case study (2005) WSEAS Transactions on Computers, 4 (1), pp. 58-63.

Kashdan, L., Conner Seepersad, C., Haberman, M. and Wilson, P. (2012) "Design, Fabrication, And Evaluation Of Negative Stiffness Elements Using SLS". *Rapid Prototyping Journal* 18 (3), 194-200

Lentejas Jr., R. (2014) *Emerging 3D Printing Markets: Top 4 | Inside3dp* [online] available from <http://www.inside3dp.com/four-countries-emerging-3d-printing-giants/> [25 April 2016]

Pierrakakis, K., Gkritzali, C., Kandias, M. and Gritzalis, D. (2015) "3D Printing: A Paradigm Shift In Political Economy". in *65Th International Studies Association's Annual Convention*. held 2015 at New Orleans. International Studies Association

Popular 3D Printers. (2013) *Selective Laser Melting (SLM)* [online] available from <http://www.popular3dprinters.com/selective-laser-melting-slm/> [14 April 2016]

Reprap. (2012) *Fused Filament Fabrication - Reprapwiki* [online] available from <http://reprap.org/wiki/Fused_filament_fabrication> [15 April 2016]

Stratasys. (2016) *Absplus-P430* [online] available from <http://usglobalimages.stratasys.com/Main/Files/Material_Spec_Sheets/MSS_FDM_ABSplusP430.pdf> [14 February 2016]

Suchy, I. (2006) *Handbook of Die Design*. New York: McGraw-Hill

Tovey, A. (2015) *Why 3D Printing Is Set To Revolutionise Manufacturing* [online] available from <http://www.telegraph.co.uk/finance/newsbysector/industry/engineering/11455696/Why-3D-printing-is-set-to-revolutionise-manufacturing.html> [25 April 2016]

Winter, R., Cotton, M., Harris, E., Maw, J., Chapman, D., Eakins, D. and McShane, G. (2014) "Plate-Impact Loading Of Cellular Structures Formed By Selective Laser Melting". *Modelling and Simulation in Materials Science and Engineering* 22 (2), 025021

Wikipedia. (2016) *Fused Deposition Modeling* [online] available from <https://en.wikipedia.org/wiki/Fused_deposition_modeling> [24 February 2016]

Wikipedia. (2016) *Selective Laser Melting* [online] available from <https://en.wikipedia.org/wiki/Selective_laser_melting> [18 February 2016]

Wohlers, T. (2014) *Wohlers Report 2014*. Fort Collins, Colorado: Wohlers Associates

Appendix

The following are the results obtained using the material ABSplus-P430

Original Model		
Force (N)	Stress (MPa)	Displacement (mm)
100	25.493	0.51748
200	49.807	0.98512
350	63.694	1.6143
500	60.121	2.1923
600	71.035	2.5616
700	81.868	2.9272
850	98.1	3.5377
1000	114.75	4.3415

Table 5: Results in original model for respective applied forces

Thickened Original Model		
Force (N)	Stress (MPa)	Displacement (mm)
64	16.03	0.31786
128	31.654	0.61525
224	54.327	1.0293
320	57.103	1.4102
384	66.475	1.6519
448	68.288	1.8856
544	63.794	2.2223
640	73.943	2.5457
704	80.613	2.7559
768	87.218	2.963
864	97.013	3.2819

960	106.67	3.6324
1024	113.01	3.8843
1088	119.25	4.1545
1184	128.31	4.6329
1280	136.7	5.2643
1344	141.64	5.8708
1372.8	155.28	6.2586
1401.6	170.42	6.6959
1430.4	186.81	7.193
1473.6	210.45	7.9736
1538.4	241.41	9.6202
1600	267.48	11.76

Table 6: Results in thickened original model for respective applied forces

Hexagon 1 Structure		
Force (N)	Stress (MPa)	Displacement (mm)
100	9.9752	0.12649
200	20.28	0.25525
350	36.343	0.45298
500	53.105	0.65681
600	64.649	0.79615
700	67.534	0.93872
850	59.59	1.1602
1000	63.364	1.3956
1100	66.959	1.5612
1200	74.035	1.733
1350	85.014	2.0025
1500	96.405	2.287
1600	104.24	2.4865
1700	112.25	2.6948
1850	124.63	3.0263
2000	137.5	3.385
2100	146.37	3.6474
2200	155.54	3.9308
2350	170.24	4.3929
2500	185.25	4.9001
2600	195.15	5.2588
2700	205.28	5.6304
2850	224.95	6.2033
3000	242.87	6.8075

Table 7: Results in hexagon 1 and hexagon 2 structures for respective applied forces

Honeycomb Structure		
Force (N)	Stress (MPa)	Displacement (mm)

10	2.1183	0.11587
20	4.309	0.23399
35	7.7577	0.41593
50	11.419	0.60453
60	13.991	0.73464
70	16.8	0.86889
85	21.695	1.0797
100	27.468	1.3053
108	31.078	1.4337
116	35.249	1.5698
128	43.339	1.7955
140	49.157	2.0838

Table 8: Results in honeycomb structures for respective applied forces

The following are the results obtained using the material ABSplus-P430

Original Model		
Force (N)	Stress (MPa)	Displacement (mm)
0	0	0
100	25.992	6.71E-03
200	51.969	1.34E-02
300	77.932	2.01E-02
400	103.88	2.68E-02
500	129.81	3.35E-02
600	155.73	4.01E-02
700	181.63	4.68E-02
800	207.52	5.34E-02
900	233.39	6.01E-02
1000	259.24	6.67E-02

Table 9: Results in original model for respective applied forces (316L SS)

Thickened Original Model		
Force (N)	Stress (MPa)	Displacement (mm)
0	0	0
160	40.375	1.01E-02
320	80.72	2.02E-02
480	121.04	3.02E-02
640	161.32	4.02E-02
800	201.58	5.02E-02
960	241.8	6.02E-02
1120	282	7.02E-02

1280	322.16	8.01E-02
1440	362.3	9.01E-02
1600	402.4	1.00E-01

Table 10: Results in thickened original model for respective applied forces (316L SS)

Hexagon Structure 2		
Force (N)	Stress (MPa)	Displacement (mm)
0	0	0
125	13.3	2.46E-03
250	26.604	4.92E-03
375	39.911	7.38E-03
500	53.222	9.84E-03
625	66.537	1.23E-02
750	79.855	1.48E-02
875	93.177	1.72E-02
1000	106.5	1.97E-02
1125	119.83	2.22E-02
1250	133.16	2.46E-02
1375	146.5	2.71E-02
1500	159.84	2.95E-02
1625	173.18	3.20E-02
1750	186.53	3.45E-02
1875	199.88	3.69E-02
2000	213.24	3.94E-02
2125	226.59	4.19E-02
2250	239.95	4.43E-02
2375	253.32	4.68E-02
2500	266.69	4.92E-02

Table 11: Results for Hexagon 1 and Hexagon 2 structure model for respective applied forces (316L SS)

Honeycomb Structure

Force (N)	Stress (MPa)	Displacement (mm)
0	0	0
10	2.1084	1.41E-03
20	4.217	2.81E-03
30	6.3258	4.22E-03
40	8.4349	5.63E-03
50	10.544	7.04E-03
60	12.653	8.44E-03
70	14.763	9.85E-03
80	16.873	1.13E-02
90	18.983	1.27E-02
100	21.093	1.41E-02
108	22.781	1.52E-02
116	24.47	1.63E-02
124	26.158	1.75E-02
132	27.847	1.86E-02
140	29.536	1.97E-02

Table 12: Results in honeycomb structure for respective applied forces (316L SS)

Model	Mass (ABSplus-P430) (gm)	Mass (316L SS) (gm)
Original Model	5.61	43.15
Thickened Original Model	6.08	46.75
Hexagon 1 Structure	6.43	49.47
Hexagon 2 Structure	7.05	54.25
Honeycomb Structure	7.38	56.88

Table 13: Mass of the models

www.ingramcontent.com/pod-product-compliance
Lightning Source LLC
Chambersburg PA
CBHW040811200526
45159CB00022B/249